ASTRONOMY AND
ASTROPHYSICS LIBRARY

Series Editors: M. Harwit, R. Kippenhahn, J.-P. Zahn

D0081338

ASTRONOMY AND ASTROPHYSICS LIBRARY

Series Editors: M. Harwit, R. Kippenhahn, J.-P. Zahn

Martin Harwit

Astrophysical Concepts

Second Edition

With 175 Illustrations

Springer-Verlag

New York Berlin Heidelberg London Paris
Tokyo Hong Kong Barcelona Budapest

Martin Harwit, National Air and Space Museum, Smithsonian Institution, Washington, D.C. 20560, USA

Series Editors

Martin Harwit	**Rudolf Kippenhahn**	**Jean-Paul Zahn**
National Air and Space Museum	Max-Planck-Institut für	Université Paul Sabatier
Smithsonian Institution	Physik und Astrophysik	Observatoires du Pic-du-Midi
Washington, D.C. 20560, USA	Institut für Astrophysik	et de Toulouse
	Karl-Schwarzschild-Straße 1	14, Avenue Edouard-Belin
	D-8046 Garching,	F-31400 Toulouse, France
	Fed. Rep. of Germany	

Cover photo: M51, Whirlpool Galaxy. Courtesy U.S. Naval Observatory.

Library of Congress Cataloging-in-Publication Data
Harwit, Martin, 1931–
    Astrophysical concepts.
    (Astronomy and astrophysics library)
    Bibliography: p.
    Includes index.
    1. Astrophysics. I. Title. II. Series.
QB461.H37 1988   523.01   87-32387

The first edition was published in 1973 by John Wiley and Sons, Inc., and was reprinted and published in 1982 by Concepts.

Typeset by Asco Trade Typesetting Ltd., Hong Kong.
Printed and bound by Edwards Brothers, Inc., Ann Arbor, Michigan.
Printed in the United States of America.

9 8 7 6 5 4 3

ISBN 0-387-96683-8 Springer-Verlag New York Berlin Heidelberg
ISBN 3-540-96683-8 Springer-Verlag Berlin Heidelberg New York

Preface to the Second Edition

My principal aim in writing this book was to present a wide range of astrophysical topics in sufficient depth to give the reader a general quantitative understanding of the subject. The book outlines cosmic events but does not portray them in detail—it provides a series of astrophysical sketches. I think this approach befits the present uncertainties and changing views in astrophysics.

The material is based on notes I prepared for a course aimed at seniors and beginning graduate students in physics and astronomy at Cornell. This course defined the level at which the book is written.

For readers who are versed in physics but are unfamiliar with astronomical terminology, Appendix A is included. It gives a brief background of astronomical concepts and should be read before starting the main text.

The first few chapters outline the scope of modern astrophysics and deal with elementary problems concerning the size and mass of cosmic objects. However, it soon becomes apparent that a broad foundation in physics is needed to proceed. This base is developed in Chapters 4 to 7 by using, as examples, specific astronomical situations. Chapters 8 to 10 enlarge on the topics first outlined in Chapter 1 and show how we can obtain quantitative insights into the structure and evolution of stars, the dynamics of cosmic gases, and the large-scale behavior of the universe. Chapter 11 discusses life in the universe, while Chapter 12 attempts a synthesis of everything discussed before, by tracing the history of the universe from its beginnings to the formation of the sun and the planets.

Throughout the book I emphasize astrophysical concepts. This means that objects such as asteroids, stars, supernovae, or quasars are not described in individual chapters or sections. Instead, they are mentioned throughout the text whenever relevant physical principles are discussed. Thus the common features of many astronomical situations are underlined, but there is a partition of information about specific astronomical objects. For example, different aspects of neutron stars and pulsars are discussed in Chapters 5, 6, 8, Appendix A, and elsewhere. To compensate for this treatment, a comprehensive index is included.

I have sketched no more than the outlines of several traditional astronomical topics, such as the theories of radiative transfer, stellar atmospheres, and polytropic gas spheres, because a complete presentation would have required extensive mathematical development to be genuinely useful. However, the main physical concepts of these subjects are worked into the text, often as remarks without specific mention. In addition I refer, where appropriate, to other sources that treat these topics in greater detail.

The list of references is designed for readers who wish to cover any given area in greater depth. I have cited only authors who actively have contributed to a field and whose views bring the reader closer to the subject. Although some of the cited articles are popular, the writing is accurate.

In this Second Edition, I have tried to update the book throughout, revising at innumerable points, page by page, to make up for the passage of 15 years since the book was first published. I have also added a totally new final chapter, which attempts to trace our current overview of how various features of the universe, large and small, are structurally related.

A book that covers a major portion of astrophysics must be guided by the many excellent monographs and review articles that exist today. It is impossible to acknowledge all of them properly and to give credit to the astrophysicists whose viewpoints strongly influenced my writing. I am also grateful for the many suggestions for improvements offered by my colleagues at Cornell and by several generations of students who saw this book evolve from a series of informal lecture notes. Finally, I would like to thank Barbara L. Boettcher for preparing the drawings.

Martin Harwit

Contents

CHAPTER 3. DYNAMICS AND MASSES OF ASTRONOMICAL BODIES

CHAPTER 4. RANDOM PROCESSES

CHAPTER 6.   ELECTROMAGNETIC PROCESSES IN SPACE   190

1

An Approach to Astrophysics

In a sense each of us has been inside a star; in a sense each of us has been in the vast empty spaces between the stars; and—if the universe ever had a beginning— each of us was there!

Every molecule in our bodies contains matter that once was subjected to the tremendous temperatures and pressures at the center of a star. This is where the iron in our red blood cells originated. The oxygen we breathe, the carbon and nitrogen in our tissues, and the calcium in our bones, also were formed through the fusion of smaller atoms at the center of a star.

Terrestrial ores containing uranium, plutonium, lead, and many other massive atoms must have been formed in a supernova explosion—the self-destruction of a star in which a sun's mass is hurled into space at huge velocity. In fact, most of the matter on earth and in our bodies must have gone through such a cata-strophic event!

The elements lithium, beryllium, and boron, which we find in traces on earth, seem to have originated through cosmic ray bombardment in interstellar space. At that epoch the earth we now walk on was distributed so tenuously that a gram of soil would have occupied a volume the size of the entire planet.

To account for the deuterium, the heavy hydrogen isotope found on earth, we may have to go back to a cosmic explosion signifying the birth of the entire universe.

How do we know all this? And how sure are we of this knowledge?

This book was written to answer such questions and to provide a means for making astrophysical judgments.

We are just beginning a long and exciting journey into the universe. There is much to be learned, much to be discarded, and much to be revised. We have excel-lent theories, but theories are guides for understanding the truth. They are not truth itself. We must therefore continually revise them if they are to keep leading us in the right direction.

In going through the book, just as in devising new theories, we will find ourselves baffled by choices between the real and the apparent. We will have to learn that it may still be too early to make such choices, that reality in astrophysics has often been short-lived, and that—disturbing though it would be—we may some day have to reconcile ourselves to the realization that our theories had recognized only superficial effects—not the deeper, truly motivating, factors.

We may therefore do well to avoid an immediate preoccupation with astrophysical "reality." We should take the longer view and look closely at those physical concepts likely to play a role in the future evolution of our understanding. We may reason this way:

The development of astrophysics in the last few decades has been revolutionary. We have discarded what had appeared to be our most reliable theories, replaced them, and frequently found even the replacements lacking. The only constant in this revolution has been the pool of astrophysical concepts. It has not substantially changed, and it has provided a continuing source of material for our evolving theories.

This pool contained the neutron stars 35 years before their discovery, and it contained black holes three decades before astronomers started searching for them. The best investment of our efforts may lie in a deeper exploration of these concepts.

In astrophysics we often worry whether we should organize our thinking around individual objects—planets, stars, pulsars, and galaxies—or whether we should divide the subject according to physical principles common to the various astrophysical processes.

Our emphasis on concepts will make the second approach more appropriate. It will, however, also raise some problems: Much of the information about individual types of objects will be distributed throughout the book, and can be gathered only through use of the index. This leads to a certain unevenness in the presentation.

The unevenness is made even more severe by the varied mathematical treatment: No astrophysical picture is complete if we cannot assign a numerical value to its scale. In this book, we will therefore consistently aim at obtaining a rough order of magnitudes characteristic of the different phenomena. In some cases, this aim leads to no mathematical difficulties. In other problems, we will have to go through rather complex mathematical preparations before even the crudest answers emerge. The estimates of the curvature of the universe in Chapter 10 are an example of these more complex approaches.

Given these difficulties, which appear to be partly dictated by the nature of modern astrophysics, let us examine the most effective ways to use this book.

For those who have no previous background in astronomy, Appendix A may provide a good starting point. It briefly describes the astronomical objects we will study and introduces astronomical notation. This notation will be used throughout the book and is generally not defined in other chapters. Those who have previously studied astronomy will be able to start directly with the present chapter that presents the current searches going on in astrophysics—the questions that we seek to answer. Chapters 2 and 3 show that, while some of the rough dimensions of the universe can be measured by conceptually simple means, a deeper familiarity with physics is required to understand the cosmic sources of energy and the nature of cosmic evolution. The physical tools we need are therefore presented in the intermediate Chapters 4 to 7. We then gather these tools to work our way through theories of the synthesis of chemical elements mentioned right at the start of this section, the formation and evolution of stars, the processes that take place in interstellar space, the evolution of the universe, and the astrophysical setting for the origins of life.

This is an exciting, challenging venture; but we have a long way to go.

Let us start.

1:1 CHANNELS FOR ASTRONOMICAL INFORMATION

Imagine a planet inhabited by a blind civilization. One day an inventor discovers an instrument sensitive to visible light and this device is found to be useful for many purposes, particularly for astronomy.

Human beings can see light and we would expect to have a big headstart in astronomy compared to any civilization that was just discovering methods for detecting visible radiation. Think then of an even more advanced culture that could detect not only visible light but also all other electromagnetic radiation and that had telescopes and detectors sensitive to *cosmic rays, neutrinos,* and *gravitational waves.* Clearly that civilization's knowledge of astronomy could be far greater than ours.

Four entirely independent channels are known to exist by means of which information can reach us from distant parts of the universe.

(a) Electromagnetic radiation: γ-rays, X-rays, ultraviolet, visible, infrared, and radio waves.

(b) Cosmic ray particles: These comprise high energy electrons, protons, and heavier nuclei as well as the (unstable) neutrons and mesons. Some cosmic ray particles consist of antimatter.

(c) Neutrinos and antineutrinos: There are two known types of neutrinos and antineutrinos; those associated with electrons and others associated with μ-mesons. A third type associated with τ-mesons is being sought.

(d) Gravitational waves.

1:1

Most of us are familiar with channel *a*, currently the channel through which we obtain the bulk of astronomical information. However, let us briefly describe channels (b), (c), and (d).

(b) There are fundamental differences between cosmic ray particles and the other three information carriers: (i) cosmic ray particles move at very nearly the speed of light, while electromagnetic and gravitational waves—as well as neutrinos, if they are found to have no rest mass—move at precisely the speed of light; (ii) cosmic rays have a positive rest mass; and (iii) when electrically charged the particles can be deflected by cosmic magnetic fields so that the direction from which a cosmic ray particle arrives at the earth often is not readily related to the actual direction of the source.

Cosmic ray astronomy is far more advanced than either neutrino or gravitational wave work. Detectors and detector arrays exist, but the technical difficulties still are great. Nonetheless, through cosmic ray studies we hope to learn a great deal about the chemistry of the universe on a large scale and we hope, eventually, to single out regions of the universe in which as yet unknown, grandiose accelerators produce these highly energetic particles. We do not yet know how or where the cosmic ray particles gain their high energies; we merely make guesses, expressed in the form of different theories on the origin of cosmic rays (Ro64a, Go69, Gu69, Hi84).

(c) Neutrinos, have zero or at least low rest mass. They have one great advantage in that they can traverse great depths of matter without being absorbed. Neutrino astronomy could give us a direct look at the interior of stars, much as X-rays can be used to examine a metal block for internal flaws or a medical patient for lung ailments. Neutrinos could also convey information about past ages of the universe because, except for a systematic energy loss due to the expansion of the universe, the neutrinos are preserved in almost unmodified form over many aeons.* Much of the history of the universe must be recorded in the ambient neutrino flux, but so far we do not know how to tap this information (We62).

A first serious search for solar neutrinos has been conducted and has shown that there are fewer emitted neutrinos than had been predicted (Da68). This has led to a re-examination of theories on the nuclear reactions taking place in the interior of the sun. First direct evidence for copious generation of neutrinos in the explosion of a supernova has given neutrino astronomy a promising new start (Hi87, Bi87).

(d) Gravitational waves, when reliably detected, will yield information on the motion of very massive bodies. Gravitational waves have not yet been directly detected, though their existence is indirectly inferred from observations on closely spaced pairs of compact stars and changes in their orbital motions about a common

---

* One aeon = 10^9 y.

center of mass. We seem therefore to be on the threshold of important discoveries that are sure to have a significant influence on astronomy (We70).

It is clear that astronomy cannot be complete until techniques are developed to detect all of the four principal means by which information can reach us. Until that time astrophysical theories must remain provisory.

Not only must we be able to detect these information carriers, but we will also have to develop detectors that cover the entire spectral range for each type of carrier. The importance of this is shown by the great contribution made by radio astronomy. Until two or three decades ago, all our astronomical information was obtained in the visible, near infrared, or near ultraviolet regions; no one at that time suspected that a wealth of information was available in the radio, infrared, X-ray or gamma-ray spectrum. Yet, today the only complete maps we have of our own Galaxy lie in these spectral ranges. They show, respectively, the distributions of pulsars and molecular, atomic or ionized gas; clouds of dust; bright, hot X-ray emitting stars and X-ray binaries; and gamma rays produced through the interaction of cosmic rays with gas in Galactic clouds.

Just as we have made our first astrophysically significant neutrino observations and are reaching for gravitational wave detection, a variety of new carriers of information have been proposed. We now speak of axions, photinos, magnetic monopoles, tachyons, and other carriers of information which—should they exist—could serve as further channels for communication through which we could gather astrophysical information. All these hypothesized entities arise from an extension of known theory into domains where we still lack experimental data. Theoretically, they are plausible, but there is no evidence that they exist in nature. Axions, photinos, and magnetic monopoles could, however, be making themselves felt through their gravitational attraction, even though otherwise unobserved. We infer from the hot, massive gaseous haloes around giant elliptical galaxies that these galaxies contain far more matter than is observed in stars and interstellar gases. The same inference is drawn from the surprisingly high speeds at which stars orbit the centers of spiral galaxies even when located at the extreme periphery of the galaxies' disks. Could this dark matter consist of such exotic particles? Tachyons, in turn, are interesting because they would travel at speeds exceeding the speed of light. Should intelligent life exist elsewhere in the universe, tachyons might provide a preferred method of rapid communication.

1:2 X-RAY ASTRONOMY: DEVELOPMENT OF A NEW FIELD

The development of a new branch of astronomy often follows a general pattern: Vague theoretical thinking tells us that no new development is to be expected at all. Consequently, it is not until some chance observation focuses attention onto a new area that serious preliminary measurements are undertaken. Many of these initial findings later have to be discarded as techniques improve.

These awkward developmental stages are always exciting; let us outline the evolution of X-ray astronomy, as an example, to convey the sense of advances that should take place in astronomy and astrophysics in the next few years.

Until 1962 only solar X-ray emission had been observed. This flux is so weak that no one expected a large X-ray flux from sources outside the solar system. Then, in June 1962, R. Giacconi, H. Gursky, and F. Paolini of the American Science and Engineering Corporation (ASE) and B. Rossi of M.I.T. (Gi62) flew a set of large area Geiger counters in an Aerobee rocket. The increased area of these counters was designed to permit detection of X-rays scattered by the moon, but originating from the sun. The counters were sensitive in the wavelength region from 2 to 8 Å.

No lunar X-ray flux could be detected. However, a source of X-rays was discovered in a part of the sky not far from the center of the Galaxy and a diffuse background flux of X-ray counts was evident from all portions of the sky. Various arguments showed that this flux probably was not emitted in the outer layers of the earth's atmosphere, and therefore should be cosmic in origin. Later flights by the same group verified their first results.

At this point a team of researchers at the U.S. Naval Research Laboratory became interested. They had experience with solar X-ray observations and were able to construct an X-ray counter some 10 times more sensitive than that flown by Giacconi's group. Instead of the very wide field of view used by that group, the NRL team limited their field of view to 10 degrees of arc so that their map of the sky could show somewhat finer detail (Bo64a).

An extremely powerful source was located in the constellation Scorpius about 20 degrees of arc from the Galactic center. At first this source remained unidentified. Photographic plates showed no unusual objects in that part of the sky. The NRL group also discovered a second source, some eight times weaker than the Scorpio source. This was identified as the Crab Nebula, a remnant of a supernova explosion observed by Chinese astronomers in 1054 A.D. The NRL team, whose members were Bowyer, Byram, Chubb, and Friedman, believed that these two sources accounted for most of the emission observed by Giaconni's group.

Many explanations were advanced about the possible nature of these sources. Arguments were given in favor of emission by a new breed of highly dense stars whose cores consisted of neutrons. Other theories suggested that the emission might come from extremely hot interstellar gas clouds. No decision could be made on the basis of observations because none of the apparatus flown had fine enough angular resolving power. Nor did the NRL team expect to attain such instrumental resolving power for some years to come.

Then, early in 1964, Herbert Friedman at NRL heard that the moon would occult the Crab Nebula some seven weeks later. Here was a great opportunity to test whether at least one cosmic X-ray source was extended or stellar. For, as the edge of the moon passes over a well-defined point source, all the radiation is suddenly cut off. On the other hand, a diffuse source is slowly covered as the moon moves across the celestial sphere; accordingly, the radiation should be cut off gradually.

No other lunar occultation of either the Scorpio source or the Crab Nebula was expected for many years; so the NRL group went into frenzied preparations and seven weeks later a payload was ready. The flight had to be timed to within seconds since the Aerobee rocket to be used only gave 5 minutes of useful observing time at altitude. Two possible flight times were available; one at the beginning of the eclipse, the other at the end. Because of limited flight duration it was not possible to observe both the initial immersion and subsequent egress from behind the moon.

The first flight time was set for 22:42:30 Universal Time on July 7, 1964. That time would allow the group to observe immersion of the central 2 minutes of arc of the Nebula. Launch took place within half a second of the prescribed time. At altitude, an attitude control system oriented the geiger counters. At 160 seconds after launch, the control system locked on the Crab. By 200 seconds a noticeable decrease in flux could be seen and by 330 seconds the X-ray count was down to normal background level. The slow eclipse had shown that the Crab Nebula is an extended source. One could definitely state that at least one of the cosmic X-ray sources was diffuse. Others might be due to stars. But this one was not (Bo64b).

Roughly seven weeks after this NRL flight the ASE-MIT group was also ready to test angular sizes of X-ray sources. Their experiment was more general in that any source could be viewed. Basically it made use of a collimator that had been designed by the Japanese physicist, M. Oda (Od65). This device consisted of two wire grids separated by a distance D that was large compared to the open space between wires, which was slightly less than the wire diameter d.

The principle on which this collimator works is illustrated in Fig. 1.1. When the angular diameter of the source is small compared to d/D, alternating strong and weak signals are detected as the collimator aperture is swept across the source. If $\theta \gg d/D$ virtually no change in signal strength is detected as a function of orientation.

In their first flight the MIT-ASE group found the Scorpio source to have an angular diameter small compared to $1/2°$. Two months later a second flight confirmed that the source diameter was small, in fact, less than $1/8°$. A year and a half later this group found that the source must be far smaller yet, less than $20''$ in diameter. On this flight two collimators with different wire spacing were used. This meant that the transmission peaks for the two collimators coincided only

1:2

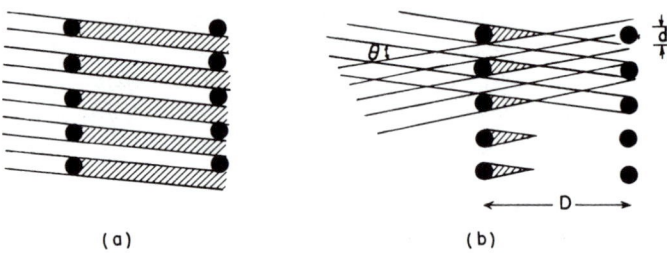

(a)   (b)

Fig. 1.1   Principle of operation of an X-ray astronomical wire-grid collimator. (*a*) For parallel light the front grid casts a sharp shadow on the rear
grid. As the collimator is rotated, light is alternately transmitted and stopped
depending on whether the shadow is cast on the wires of the rear grid or
between them. (*b*) For light from a source whose angular dimension $\theta \gg d/D$,
the shadow cast by the front grid is washed out. Rotation of the collimator
assembly then does not give rise to a strong variation of the transmitted
X-ray flux.

for normal incidence and, in this way, yielded an accurate position of the Scorpio
source (Gu66). An optical identification was then obtained at the Tokyo Observatory and subsequently confirmed at Mount Palomar (Sa66a). It showed an
intense ultraviolet object that flickered on a time scale less than one minute. These
are characteristics associated with old novae near their minimum phase.

The brightness and color of neighbouring stars in the vicinity of Sco XR-1
showed that these stars were at a distance of a few hundred light years from the
sun, and this gave us a good first estimate of the total energy output of the source.
A search on old plates showed that the mean photographic brightness of the
object had not changed much since 1896.

Interestingly, the 1969 discovery that the Crab Nebula contains a pulsar sent
X-ray astronomers back to data previously collected. Some of these records showed
the characteristic 33 millisecond pulsations, and showed that an appreciable frac-
tion of the flux—10 to 15%—comes from a point source—now believed to be a
neutron star formed in the supernova explosion (Fr69). Our views of the Crab as a
predominantly diffuse X-ray source had to be revised.

Many other Galactic X-ray sources have by now been located and identified;
and frequently they have a violet, stellar (pointlike) appearance similar to Sco XR-1.
These objects sometimes suddenly increase in brightness by many magnitudes
within hours. Others pulsate regularly, somewhat like the Crab Nebula pulsar.
The range of X-ray energies at which the observations have by now been carried
out, is quite wide too, and both visual and X-ray spectra are available for many
sources.

1:2

By now, hundreds of extragalactic X-ray sources, many of them quasars or galaxies exhibiting violently active nuclei, have also been seen. The first of these was M87, a galaxy known to be a bright radio source (By67). It is a peculiar galaxy consisting of a spherical distribution of stars from which a jet of gas seems to be ejected. The jet is bluish in visible light, and probably gives off light by virtue of highly relativistic electrons spiraling about magnetic lines of force and emitting radiation by the synchrotron mechanism (see Chapter 6)—a mechanism by which highly accelerated energetic particles lose energy in a synchroton.

Theorists have now proposed a variety of explanations for cosmic X-ray sources and also for the continuum X-ray background that appears to permeate the universe. Many experiments are being planned to test these theories. X-ray, visual, infrared, and radio astronomers compare their results to see if a common explanation can be found. Progress is rapid and perhaps in a few years this field will no longer be quite as exciting. But by then another branch of astronomy will have opened up and the excitement will be renewed.

The fundamental nature of astrophysical discoveries being made—or remaining to be made—leaves little room for doubt that a large part of current theory will be drastically revised over the next decades. Much of what is known today must be regarded as tentative and all parts of the field have to be viewed with healthy skepticism.

We expect that much will still be learned using the methods that have been so successful in the past. However, there are parts of astrophysics—notably cosmology—in which the very way in which we think and our whole way of approaching scientific problems may be a hindrance. It is therefore useful to describe the starting point from which we always embark.

1:3 THE APPROPRIATE SET OF PHYSICAL LAWS

Nowadays *astrophysics* and *astronomy* have come to mean almost the same thing. In earlier days it was not clear at all that the study of stars had anything in common with physics. But physical explanations for the observations not only of stars, but of interstellar matter and of processes that take place on the scale of galaxies, have been so successful that we confidently assume all astronomical processes to be subject to physical reasoning.

Several points must, however, be kept in mind. First, the laws of physics that we apply to astrophysical processes are largely based on experiments that we can carry out with equipment in a very confined range of sizes. For example, we measure the speed of light over regions that maximally have dimensions of the order of 10^{14} cm, the size of the inner solar system. Our knowledge of large-scale dynamics also is based on detailed studies of the solar system. We then extrapolate the dynamical laws gained on such a small scale to processes that go on, on a cosmic scale of $\sim 10^{18}$ to 10^{28} cm; but we have no guarantee that this extrapolation is warranted.

It may well be true that these local laws do in fact hold over the entire range of cosmic mass and distance scales; but we only have to recall that the laws of quantum mechanics, which hold on a scale of 10^{-8} cm, are quite different from the laws we would have expected on the basis of classical measurements carried out with objects 1 cm in size.

A second point, similar in vein, is the question of the constancy of the "constants of nature." We do not know, in observing a distant galaxy from which light has traveled many *aeons*, whether the electrons and atomic nuclei carried the same charge in the past as they do now. If the charge was different, then perhaps the energy of the emitted light would be different too, and our interpretation of the observed spectra would have to be changed.

A third point concerns the uniqueness of the universe.

Normal questions of physics are answered by experiment. We alter one feature of our apparatus and note the effect on another. Cosmic questions, however, do not permit this kind of approach. The universe is unique. We cannot alter phenomena on a very large scale, at least not at our current level of technological development; and if we did, it is not clear that we would be able to discern real changes. There would simply be no available apparatus that would not in itself become affected by the experiment—no reference frame against which to detect the change. In short, we may not be asking questions that can be answered in physical terms, because the methods of physics, and more generally of all science, depend on our ability to conduct experiments; truly cosmic problems may just not permit such an approach.

This then is the current situation: We know a great deal about some as yet apparently unrelated astronomical events. We feel that an interconnection must exist, but we are not sure. Not knowing, we divide our knowledge into a number of different "areas": cosmology, galactic structure, stellar evolution, cosmic rays, and so on. We do this with misgivings, but the strategy is to seek a connection by solving individual small problems. All the time we expect to widen the areas of understanding, until some day contact is made between them and a firm bridge of knowledge is established between previously separated domains.

How far will this approach work for us? How soon will the philosophical difficulties connected with the uniqueness of the universe arise? We do not yet know; but we expect to face the problem when we get there.

In the meantime we can address ourselves to a number of concrete problems which, although still unsolved, nevertheless are expected to have solutions that can be reached using the laws of physics as we know them. Among these are questions concerning the origin and evolution of stars, of galaxies, of planetary systems. There are also questions concerning the origins of the various chemical elements;